Simple Innovations in Machine Design

Providing Manufacturers with Simple Novel
Concepts

OLUWAFUNBI SIMOLOWO

SIMPLE INNOVATIONS IN MACHINE DESIGN

DEDICATION

This book is dedicated to the all-wise, all-knowing designer and creator of the heaven and earth: The Lord God Almighty

CONTENTS

ACKNOWLEDGMENTS

I give all thanks to God Almighty the giver of all knowledge and wisdom for making the first volume of this dream-book a reality. I wish to also thank my family for their role in putting this book together. I would also like to express my appreciation to the engineering students of the Faculty of Technology; University of Ibadan whose names I have also listed in this book that worked on the group projects presented in this book

Design Contributors

Triune Cooker
Azeez Oluwafemi
Shittu Saheed
Fasae Taiwo

Ternion Cooker
Abdulsalaam Bolaji I.
Anyalibe Romanus O.
Ojelabi Adekunle S.
Alaba Oluwasegun T.
Awosanya Taiwo O.
Adedokun Kazeem O.

The Hydraulic Vise
Jimoh Oluseun
Abifade Olamide
Ipaye Musiliyu. A
Adegboye Oluwatoni
Meludu Ebelechukwu
Isoghie Edward James
Abolo Emmanuel
Sanusi Mubarak
Adedokun Iyinoluwa

Pedal hydraulic Jack
Opakunle Olakunle Charles
Olaoye Feyikemi Esther
Chukwuemeka Paul
Babatunde Ifeoluwa Paul
Eegunjobi Victor Adeolu
Adedamola Samuel
Oyelade Paul Oluwasayo

Perfostape
Dada Taiwo   O
Akano Damola   I
Adeyemi Cyril   A
Agarin Ejiroghene P
Aturu Olawale   O
Lawal Mujeeb

Poster Remover
Abdulrasaq Abdulazeez A.
Amaji, Chukwuebuka P.
Ayoade Oluseyi B.
Bankole Ruth O.
Okesola, Michael O.
Oluyemi Kolade S.
Tanimola Samuel O.

Okra Compound Grater
Banjo Oluwatimilehin
Hassan Oluwatomiwo
Uzondu Charles.C
Subair Abdulazeez
Alafe Kikelomo
Odunowo Tomisin

Dual Mixer
Namso I. Samson
Kolade Habeebullah O.
Lijoka Oluwatomisin E.
Adepoju Gideon O.
Nnate Florence N.
Omotosho Gboloahan

Two- Point Grinder
James Olusegun
Multi- plug Spanner
Adeyinka Daniel
Ogunmoye Victor
Moshood Oluwakorede
Daodu Samsudeen A.

Multi- plug Spanner
Adeyinka Daniel
Ogunmoye Victor
Moshood Oluwakorede
Daodu Samsudeen A.

PART ONE

COOKERS

[1]TRIUNE COOKER

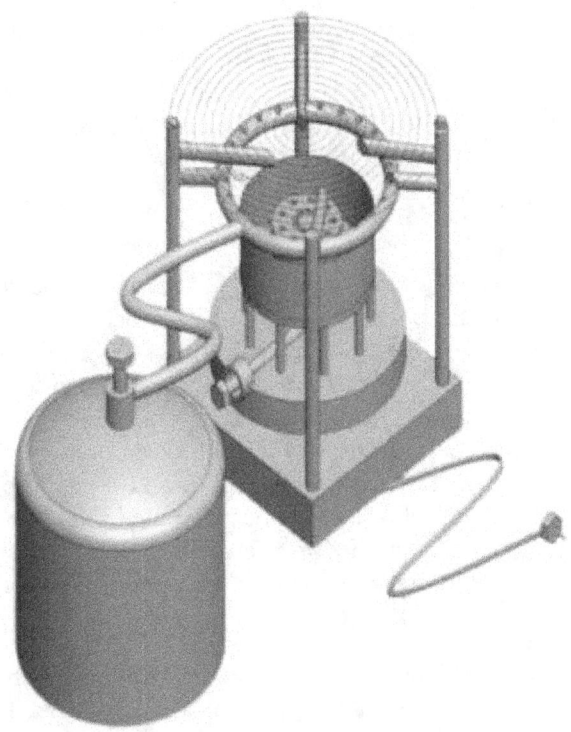

**An innovative design of a Triune cooker
that uses three sources of heat**

Inventing is the mixing of brains and materials. The more the brains you use, the less the materials you need. ---------Charles F. Kettering

1.1   Need for Design

There are three common sources of heat for cooking. They are; Kerosene (A combustible hydrocarbon liquid); Electricity (Power); Natural gases (A highly combustible hydrocarbon gas). The common problems encountered when cooking is inadequate access to these resources. More so, the erratic supply of electricity in some parts of the world accounts for the low utilization of electric cooker. The high cost and fluctuation of gas availability is also a factor to be considered. This design utilizes the three energy sources in its operation.

1.2   Description of Existing Cooker Designs

The existing designs are in three separate forms namely, kerosene stove; gas cooker and electric cooker

Kerosene Stove

Kerosene stoves are widely used and are readily available. They have high heat output, are cheaper to operate and are efficient. They are multi-fuel in design and will burn just anything that burns but gasoline should be avoided at all cost. They are simple to construct and require little to no maintenance. A kerosene wick stove (Figure 1.1) works much like a candle. One end of a fiber wick rests in a reservoir of fuel and fuel is drawn up through the wick by capillary action. The wick is moved up and down into an annular space formed by two thin walled concentric perforated steel shields (flame holders) by a lever. A flame is applied to the other end of the saturated wick, thus igniting the fuel and drawing fuel into wick, thus maintaining the flame.

Gas Cooker

The gas cooker (Figure 1.2) is a cooking device which uses natural gas, propane, butane or other flammable gas as fuel source. It consists of a gas cylinder, hose, gas dispenser and control knob. Natural gas (propane) being the purest form of hydrocarbon existing is set alight and combustion occurs unlike other form of fuel which is turned to vapor before combustion takes place. The natural gas flows from the gas cylinder through the hose to the gas dispenser by opening a valve where it comes in contact with fire thereby generating heat for cooking. They have accurate cooking temperature, evenly distributed heat, lower heat emissions and faster cooking time and do not use power.

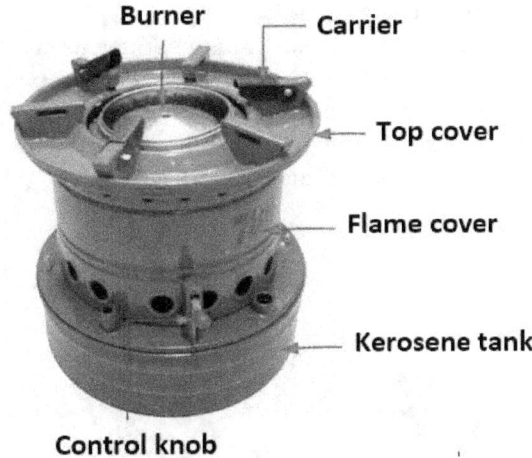

Figure 1.1: kerosene wick stove
Source: Yangzhou Kerosene Stoves Co Ltd

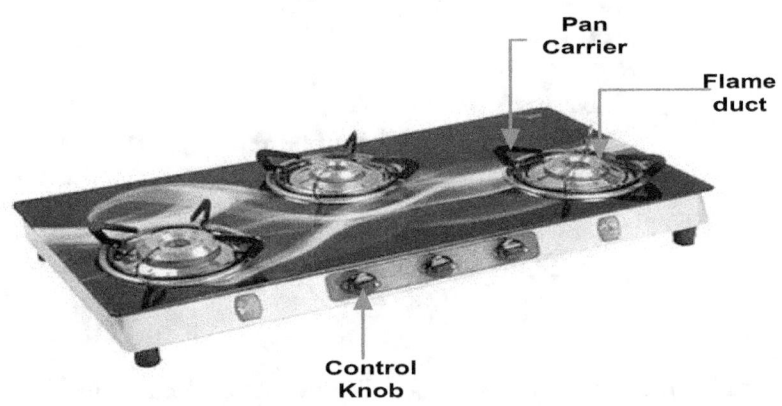

Fig. 1.2: The gas cooker
Source: Baweja Appliances - TradeIndi

Electric Cooker

The electric cooker (Figure 1.3) is an electric powered cooking device for heating and cooking of food. It consists of element, plug, control knob and connecting cable. Electric current passes through the cable and generates heat that radiates out of the heating elements. The different heat settings are controlled by either adjusting the voltage through the cable or by using a thermostat to turn the current on and off as necessary. The element converts the electrical energy to heat energy for cooking.

1.3   Design of Triune Cooker

The Triune cooker (Figs 1.4 and 1.5) is a combination of three cookers. The bottom is made up of a kerosene cooker that also serves as a carrier for the gas burners. The gas cylinder for safety purposes is located outside the system and connected to its cooker with a hose. The electric heating element is joined to the frame of the cooker at the top with a hinge.

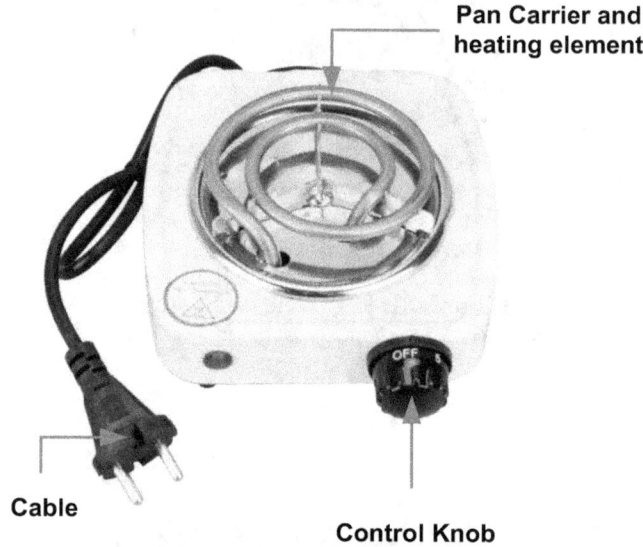

Pan Carrier and heating element

Cable

Control Knob

Fig. 1.3: Electric Cooker
Source: AliExpress Delidge

Kerosene cooker unit: The fuel is supplied by capillary action through the wick which sustains fire as fuel (kerosene) is supplied by the cylinder from which the wick comes out.

Gas cooker unit: The gas cylinder is external and is connected to its cooker through a hose. The regulator is on the gas cylinder to regulate the gas dispense rate, this is the rate that determines the level of fire produced for cooking. When the gas is dispensed, the system is lit up and the flame burns to supply the required heat for cooking. The bottom stack is connected to the top stack with a hinge.

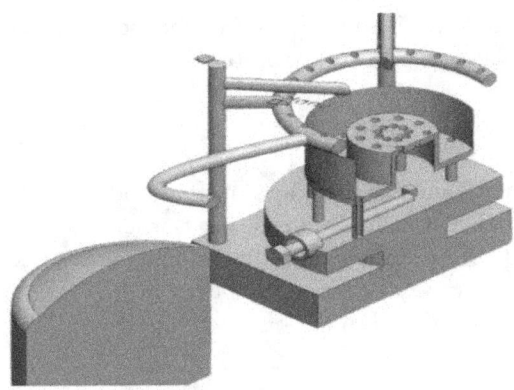

Fig. 1.4: Sectional view of Triune Cooker

Electric Cooker: The switch and the control for the electric cooker are located below the bottom stack and the wire is passed through the pole to the top stack where it is connected to the filament. The electrical energy is being conveyed to reasonable required heat by the positive and negative charges passing through the filament, where the heat is used for cooking.

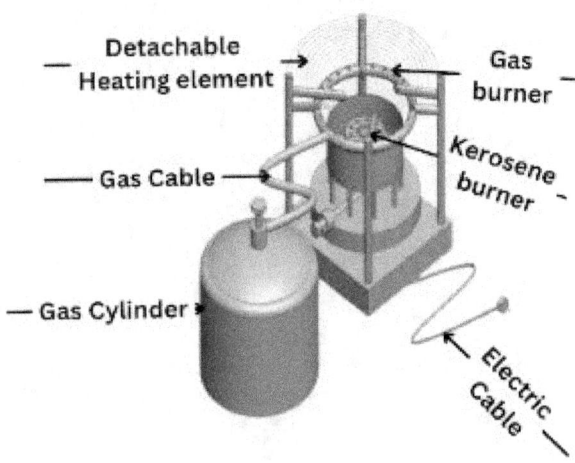

Fig. 1.5: Triune Cooker Assembly

1.4   Assembly Guidelines

1. The kerosene tank is joined to the top of the electrical box.

2. The stands are also joined to the top of the electrical box, i.e. four stands around the edges.

3. The wick stand, burner and baffles are joined respectively to the top of the kerosene tank.

4. The gas heater is joined to the edges of the stands

5. The hose joins the gas cylinder to the gas burner.

6. The hinge and the lock are fixed on to stands, each connected to a wire that carries electric current inside the stands.

7. The filament is connected to the hinge.

[2] TERNION MULTI-COOKER

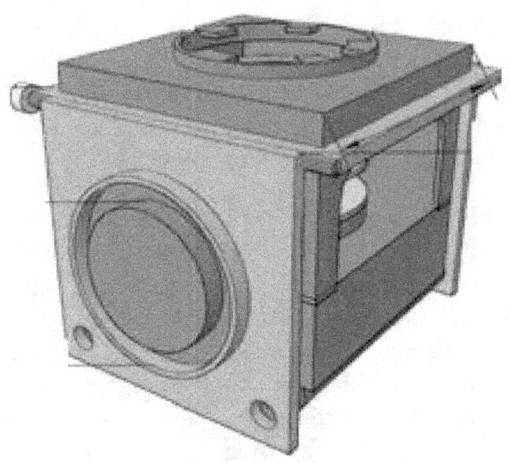

An innovative design of a compact 3-in-1 cooker

The thing that hath been, it is that which shall be; and that which is done is that which shall be done: and there is no new thing under the sun.------------Ecclesiastes 1:9 (KJB)

2.1   Description of Ternion-Multi Cooker

The Ternion multi-cooker is a combination of three different forms of cookers.

- The kerosene stove,
- The electric cooker
- The gas cooker.

They are joined together with the aid of strong metallic hinges. The kerosene stove is located at the middle of the device serving as a stand for the multi-cooker. The electric cooker and gas cooker can be folded inwards and the gas cylinder is detachable, as to avoid cumbersomeness. This design concept makes simultaneous cooking with three different energy sources possible. Ceramics have been introduced into the model to replace the conductive iron parts in order to reduce the quantity of heat produced.

2.2 Design Mechanisms

The mechanism involves the rack and pinion system which converts the rotational motion of the control knob to translational motion of the wick frame. This allows easy adjustment of the wick. When the control knob is being turned, the shaft connected to the control knob is also turned thereby creating rotational motion on the gear connected to the shaft. The rotational motion of the gear is converted to translational on the pinion. The movement of the pinion move the wick frame (housing for wick) thereby controlling the flame. Unlike the existing kerosene stove where the pinion moves above the top of wick rack when controlled to the maximum, thereby causing incomplete combustion. The pinion in stove compartment of ternion multi cooker doesn't move above the top of the wick rack. The hinge attached to the equipment moves at an angle 90°, thereby making it possible to fold for compactness and mobility.

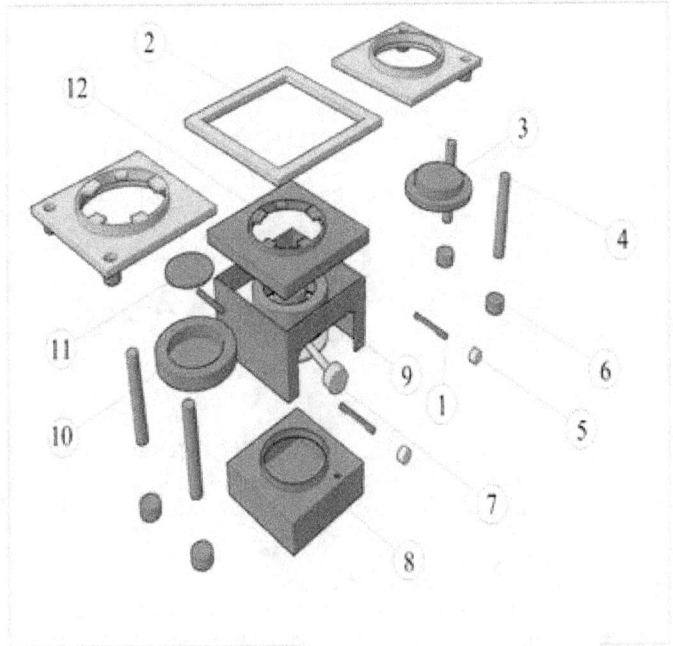

Fig. 2.1 Exploded View of Ternion-Multi Cooker

2.3   Assembly Guideline of Ternion-Multi Cooker

The assembly guidelines for the ternion-multi cooker are herein described with reference to Figure 2.1 and 2.2.

- The wick is inserted into the hole of the wick frame.
- The wick-rack is placed to sit properly on the kerosene stand.
- The combination of the burner, flame shield and flame holder is placed to sit properly on the bed provided for it on the wick-rack(7)
- The frame (2) is placed on the body
- The cover (12) is placed on the frame (2) to fit
- The gas cooker is placed on the seat provided for it on the frame
- The gas burner (11) is placed on the gas dispenser (10)
- The pan carrier is placed on the gas dispenser (10)
- The gas cooker s connected ito the gas cylinder with a hose
- The electric cooker is placed on the seat provided for it on the frame
- The rubber (6) is fixed to the stand then to the frame(2)

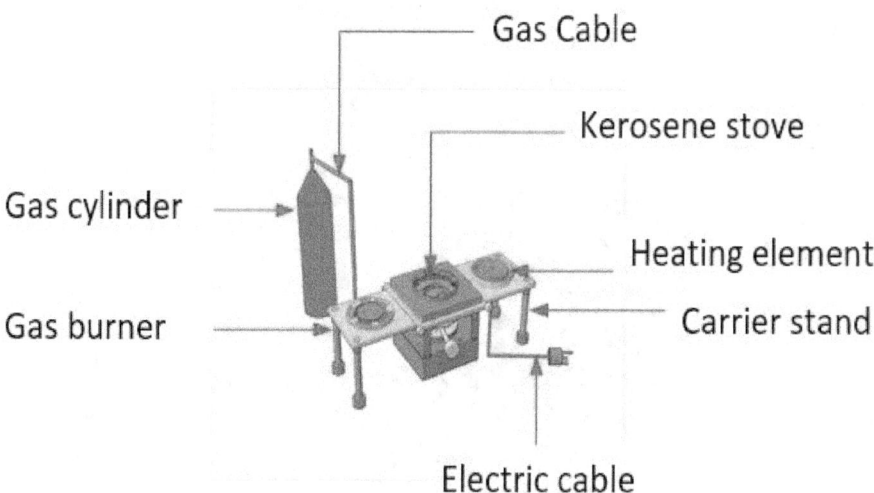

Fig 2.2: Ternion cooker assembly

2.4 Sizing of Ternion Multi Cooker

The total weight of the ternion-multi cooker is approximately 4kg. The Electric cooker is 0.6kg; the Kerosene cooker is 1.2kg and the Gas cooker is 0.8kg. The Four rods are 0.4kg. Shown in Figure 2.3 are the dimensions of the basic parts

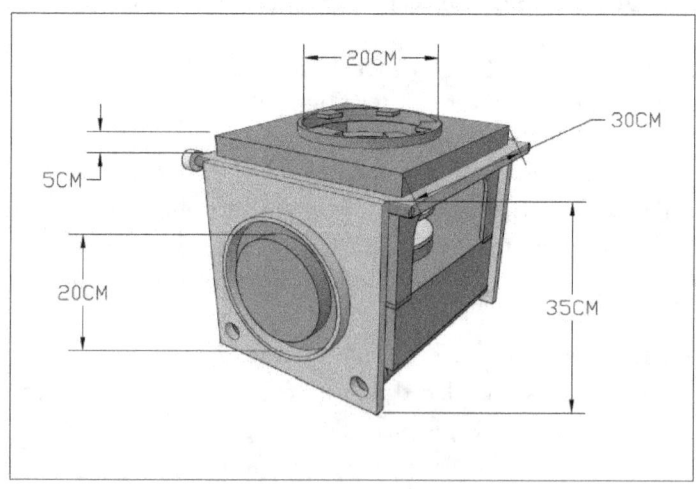

Fig. 2.3: Sizing of Basic Parts of Ternion Multi Cooker

PART TWO
HYDRAULIC DEVICES

[3] HYDRAULIC BENCH VISE

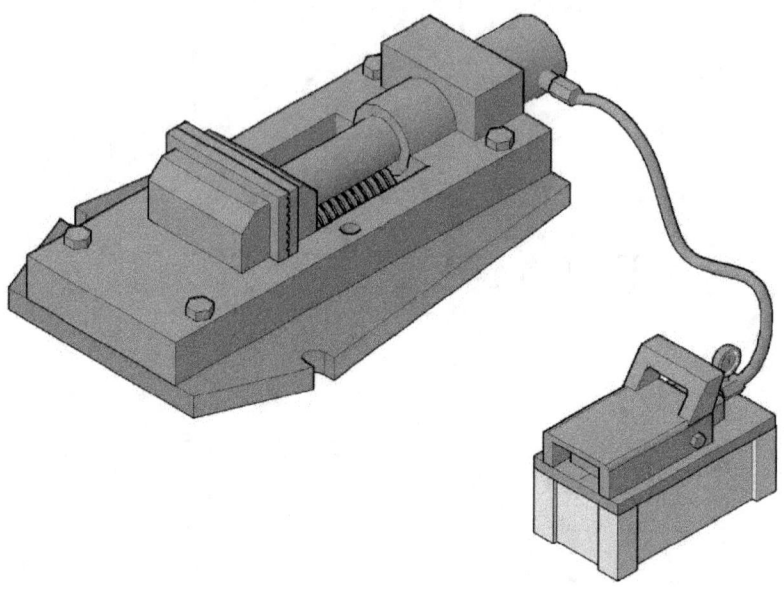

A foot-operated hydraulic bench vice

Invention is 99% perspiration and 1% inspiration. ------Thomas Alva Edison

3.1 Need for New Design

A bench vise is like an extra hand and it's a common tool found in any workshop or garage. It is attached to a work bench and its purpose is to hold a work piece steady, allowing the use of both hands to work on the material with other tools. They are ideal for sawing, sanding, planning, screwing, soldering and other workshop activities. The newly designed hydraulic bench vise is a general purpose tool that can have multiple clamping and holding applications. A hydraulic bench vise provides power and performance with the efficiency of hydraulic power. The user operates the vise by use of hydraulic foot pedal. It is user and handicap friendly and eliminates physical limitations and risk factors.

After considering the shortcomings of the existing design, we consider the need for a new design, a hydraulic bench vise. The new design has the following characteristics;

- It is durable, powerful and precise
- It reduces the high risk for injury and/or strained muscles.
- It provides dependable power, control or stability. This is even more evident because it reduces additional help which is necessary for set-up and operation.
- It is user and handicap friendly and eliminates physical limitations and risk factors.
- It allows for consistency, repeatability and perfect clamping pressure.
- It provides a hand free operation for safe and controlled actuation.

3.2 Existing Design

The existing design as shown in Figure 3.1 is the conventional bench vise which is manually operated. The operator has to manually crank down on its lever to operate it. There is a high risk of injury and strained muscles, it cannot also provide dependable power, stability and control in its set up additional help may be required.

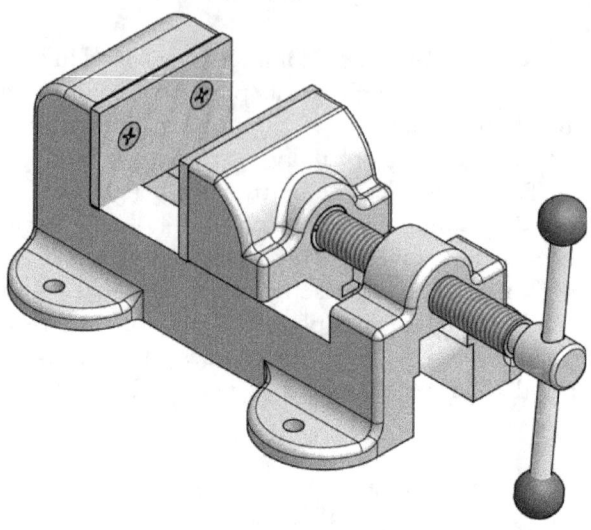

Fig. 3.1: The conventional bench vise

3.3   Description of Hydraulic Bench vice Operation

The hydraulic vise as shown in Figure 3.2 – 3.5 comprises jaws (movable [4] and fixed [3]), hydraulic cylinder [5], hydraulic hose [6], pressure gauge [8], foot pedal [9], valve [10]. When the foot pedal [9] is activated (i.e. pressed down), force is exerted on the fluid inside the hydraulic tank [7] thereby increasing the pressure on the fluid which causes the fluid to flow through the opening of the hydraulic tank [7] into the hydraulic holes [6]. The fluid flowing through the hydraulic holes enters the hydraulic cylinder [5] where pressure is exerted on the piston, the piston then moves hence moving the movable jaw [4] towards the fixed jaw [3]. The movement of the movable jaw [4] towards the fixed jaw [3] causes the clamping action of the work piece. When the force exerted on the foot pedal[9] is retracted, the tension spring[2] incorporated on the body of the vise and attached to the body of the movable jaw [4] will overcome the pressure left in the cylinder[5], making the cylinder [5] to retract faster, also a valve [10] has been introduced to keep the hydraulic fluid in place so as to ensure removal of the force on the foot pedal [9] without causing a consequent de-clamping of the work piece, when the work piece is to be removed the valve [10] can be turned which in turn allows the fluid to flow into the reservoir. A gauge [8] is also incorporated on the outlet of the reservoir which measures the pressure going into the hydraulic cylinder [5]. The base [1], basically is the part of the hydraulic vise attached to the workbench

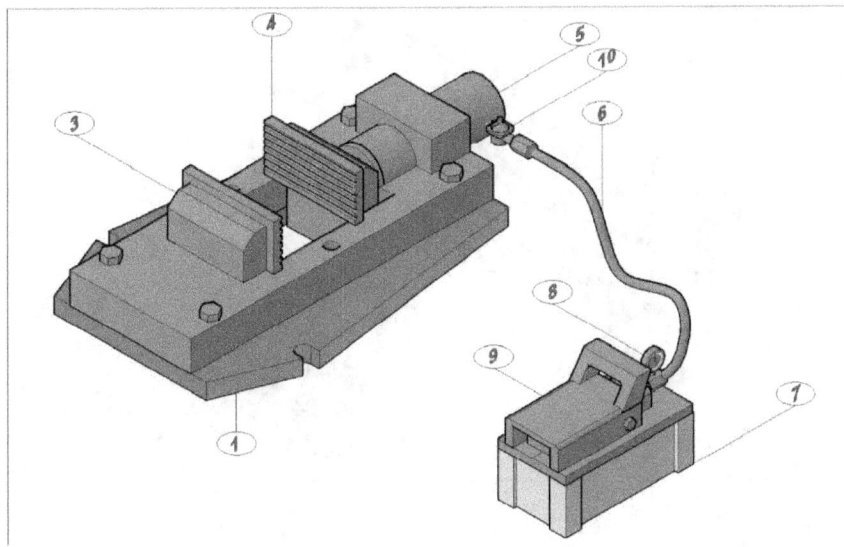

1-base; 2-collar; 3-handle ball; 4-jaw; 5-set screw; 6-slide key; 7-sliding jaw; 8-special key; 9-vise screw; 10-handle rod; 11-screw

Fig. 3.2: The Foot operated hydraulic bench vice

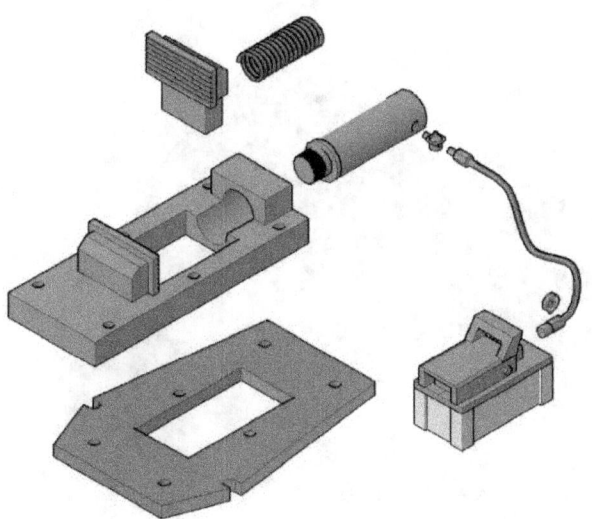

Fig. 3.3: Exploded view of hydraulic bench vise

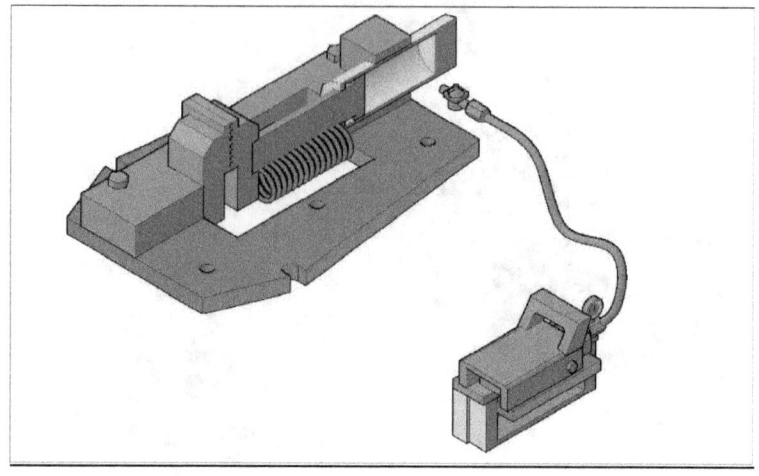

Fig. 3.4: A sectioned view of hydraulic bench vise

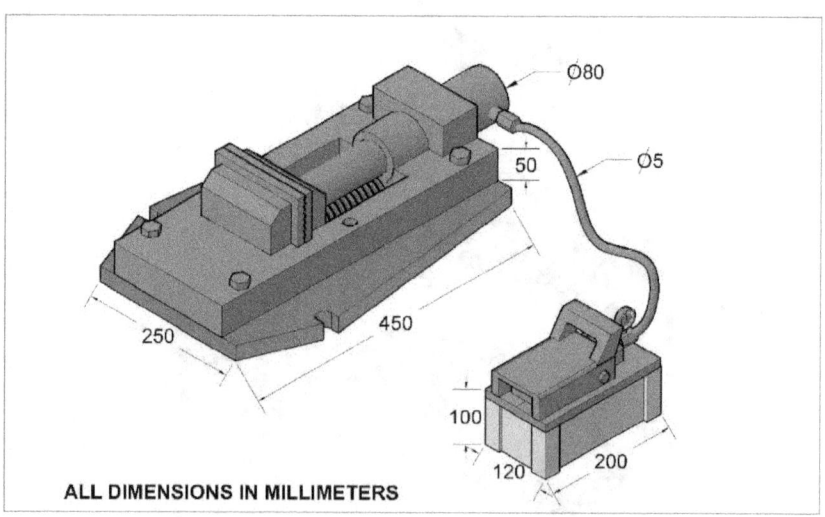

Fig. 3.5: Sizing of Basic Parts Hydraulic Bench vise

3.4   Design Mechanism

Hydraulic Cylinder: - This converts power from a pressurized hydraulic fluid which is typically oil. The hydraulic cylinder consists of a cylindrical barrel, in which a piston is connected to a piston rod which can move forth and back.

Hydraulic hose- This is a flexible hollow tube designed to carry fluids from one location to another. A hose is usually cylindrical having a circular cross-section it can also be a high pressure, thermoplastic or synthetic rubber hose that carries fluid to transmit force on exertion of pressure.

Hydraulic valve- This is a device that regulates or controls the flow of a fluid by opening, closing or partially obstructing the passage of the fluid.

Pressure gauge- This is used to measure the exerted pressure by the foot pedal from the hydraulic reservoir.

Foot pedal- This is actuated to exert force on the fluid inside the hydraulic tank by increasing the pressure in the fluid.

Spring- This is made of mild steel and it's incorporated on the vise to perform a faster/quick return stroke of the sliding jaw (Figure 3.10), it works on the principle of compression and tension.

Hydraulic Tank- This acts as a reservoir for the hydraulic fluid (oil), the hydraulic fluid is stored in this tank, it acts as the given area so when the foot pedal is actuated, Pascal's law is obeyed and the hydraulic fluid moves.

3.5   Assembly Guideline of Hydraulic bench vice

The hydraulic component must be examined for any kinks in the hose or other form of damage before use. The foot pedal is used to advance and close the vise jaws. The base of the hydraulic bench vise is made of forged (heated and beat to shape) cast iron it serves as the projection/form of the hydraulic bench vise. The Jaws are made of cast iron, they both (fixed and movable) have a jaw plate.   The base has a body frame attached/screwed to it, the body frame is then attached to the fixed jaw by means of permanent welding so as to ensure a firm attachment, the body frame was forged to contain a n hydraulic cylinder, a spring, a movable jaw and a fixed jaw.

The hydraulic cylinder consists of an assembled hydraulic barrel, piston. The piston moves forth and back when the foot pedal is actuated hence the spring is firmly attached to the piston in the hydraulic cylinder and the spring causes fast return stroke of the movable jaw. The valve is attached to the hydraulic hose and hydraulic cylinder by means of a male coupler at one end, and at the other end it is attached to the pressure gauge and the hydraulic tank (reservoir) by the male coupler. Then the wholly assembled bench vise can be attached to the table or bench as preferred by the engineer using screws ranging from (14" to 16"). (All the materials used for this design project were either purchased or fabricated from an existing design and re-modeled.)

[4] PEDAL HYDRAULIC JACK

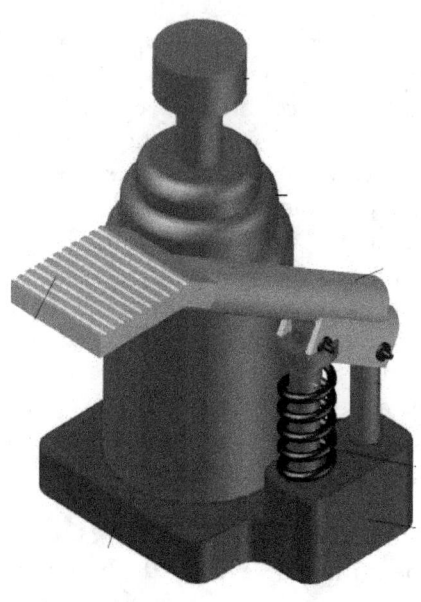

A hydraulic car jack with extendable pedal bar

Is there anything whereof it may be said, See, this is new? it hath been already of old time, which was before us-------Ecclesiastes 1:10 (KJB)

4.1 Need for New Design

The function of this new design is to raise machines but mostly used for automobiles especially trucks for easy removal of tyres or faulty parts of the vehicles. It has a leg pedal to make its application easier. The new pedaled hydraulic jack reduces inconveniences, pain and stress undergone when using the existing hydraulic jack. The new design of the pedaled hydraulic jack is needed because of the stress associated with using the existing hydraulic jack that requires the operator to bend before the jack could be used. This wastes so much time and uses up so much energy for the jacking process leaving the operators with back pains after jacking operations. Hence, the need for a jack that eliminates bending in its usage.

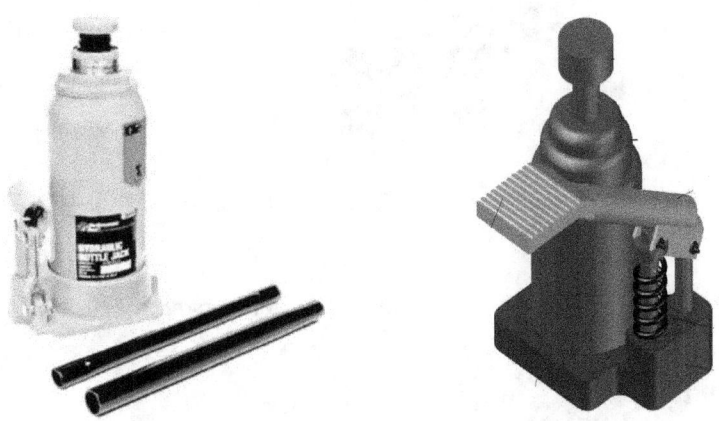

Fig. 4.1: **Difference between Existing Design and New Design**
Source: EBay

Unlike the conventional hydraulic jack, the new design has a pedal and a larger base area for better stability. A return spring is also incorporated to aid continuous pressing of the hydraulic pedal. The new design also has two rams and an extendible connecting bar to increase the length of the pedal for a more effective application when placed under vehicles.

4.2 Mechanism of New Design

The new design of the hydraulic jack works based on the principle of expansibility of springs. It is fitted with an extendible connecting bar which is connected to a pedal. The hydraulic jack is set up as shown Figures 4.2 – 4.4. On applying pressure to the pedal, it presses the spring that is directly below it. The spring thus moves down with the pedal and pushes the ram that is connected to the jack handle down in the process. The ram moves down into the oil case and thus increases the force on the oil in a specified constant cross sectional area. This force increases the pressure on the ram and thus forces the inner ram out of the ram casing. The presence of the spring now forces the pedal to go back to its initial position without any effort from the operator of the hydraulic jack

Fig. 4.2: Sectional view of Hydraulic Jack

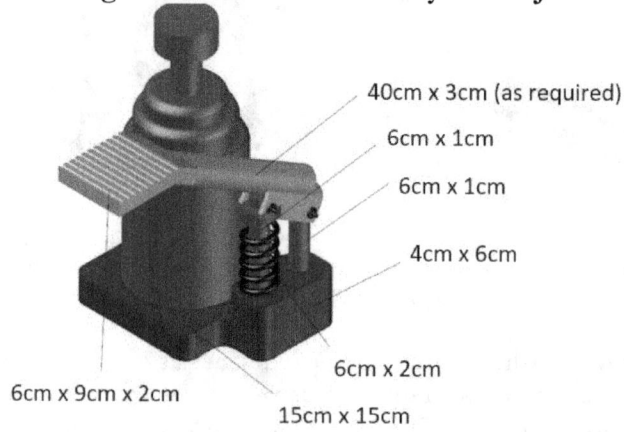

40cm x 3cm (as required)

6cm x 1cm

6cm x 1cm

4cm x 6cm

6cm x 2cm

6cm x 9cm x 2cm

15cm x 15cm

Fig. 4.3: Exploded View of Hydraulic Jack

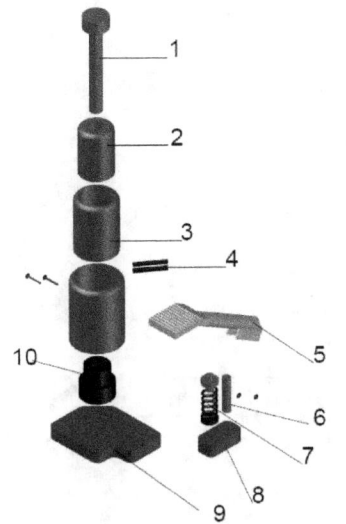

1.RAM
2.RAM CASING
3.RAM CASING
4.SPRINGS
5.JACK HANDLE
6.RAM
7.RETURN SPRING
8.OIL CASING
9.RAM BASE
10.PRESSURE CIRCULATOR

Fig. 4.4: Exploded View of Hydraulic Jack

PART THREE

SIMPLE

MECHANCAL DEVICES

[5] PERFOSTAPE

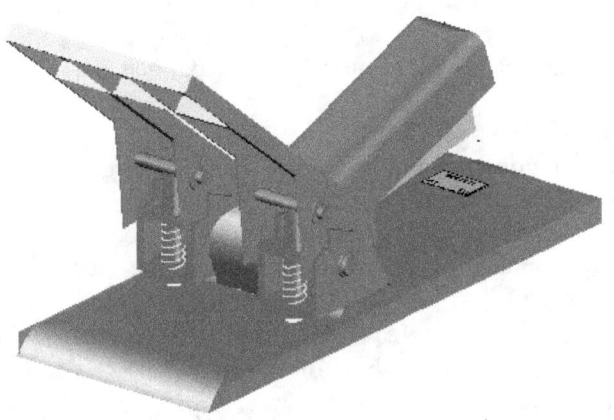

A materials-saving and cost-effective composite stapler and perforator

The ideal engineer is a composite ... He is not a scientist, he is not a mathematician, he is not a sociologist or a writer; but he may use the knowledge and techniques of any or all of these disciplines in solving engineering problems. -----------N. W. Dougherty

5.1 Need for New Design

Studies have shown that people that use the stapler on a regular basis tend to be faced with occasions where they need to use the perforator and also people that use the perforator in a similar manner, frequently find themselves in need of the stapler as well. This situation could be very tasking at times hence the idea about a device that combines the functions of both the perforator and the stapler which is named the **Perfostape**. The 'perfostape' is a mechanical device that has been designed to give the user the ability to perforate and staple using the same device unlike the ordinary stapler and perforator that are separated and work individually.

5.2 Description of Existing Design

There are two existing designs that brought about the design of the perfostape. The designs are a stapler and a perforator.

A stapler: This is a mechanical device (Figure 5.1) that joins sheets of paper or similar material by driving a thin metal staple through the sheets and folding the ends to secure the pages.

- **The base** is a flat, box-shaped container that ensures stability. Thus, by providing this stability, the base of the stapler also ensures precision, as slippage caused by the force required is prevented, and the staple is placed in the document where intended. The base also holds the other parts of the stapler in place, acting as a mount for the other parts.
- **The magazine cartridge** is a metal container that is hinged to the base and holds the staples. The magazine enables the stapler to operate more efficiently, as the magazine can hold several hundred staples, as opposed to reloading staples one by one.

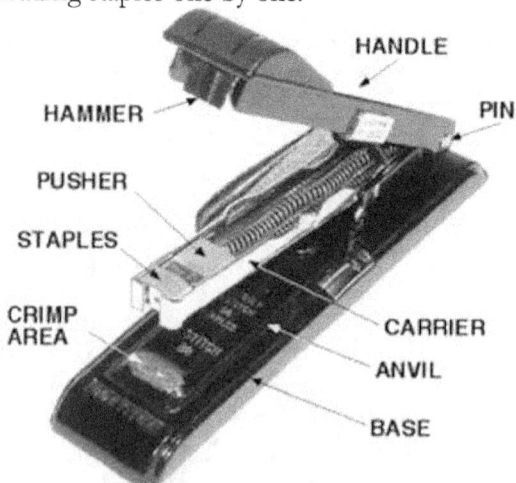

Fig. 5.1: The Stapler
Source: Zachary MacChesney

- **The carriage** is a metal rectangular box which covers the magazine cartridge. Its main purpose is "to serve as a protector to the staples "
- **The spring** is a spiral material that pushes the staple pins one after the other.
- **The tooth** is a metal material that pushes the staple pins down the hole in the magazine cartridge to be stapled.
- **The arm** is a material on which force is applied.
- **The ergonomic thumb** and finger grips are there just for easy handling.
- **The pin** is a metal material that joins all the stapler components together.

A perforator: This is a mechanical device that makes holes in materials such as papers and cards (Figure 5.2).

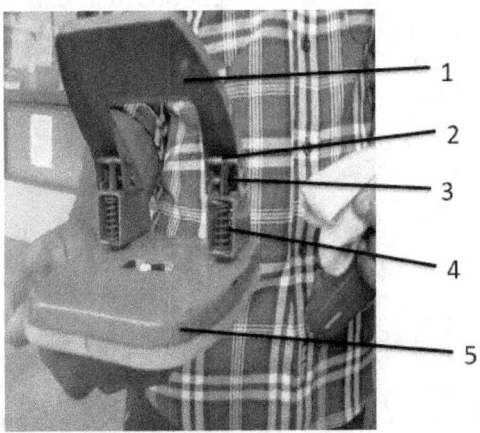

Fig. 5.2: The Perforator and User

- **Part 1** is the perforator handle on which force is applied.
- **Part 2** is the cylindrical rod which helps to push down paper punch (part 3) accurately and precisely.
- **Part 3** is the paper punch which creates the holes in the papers due to a downward motion.
- **Part 4** is the spring which holds the paper punch firmly and also helps the mechanism to return back to its initial position after perforating.
  Part 5 is the base in which the cut paper is pushed through. The paper can be removed from the base after cutting.

5.3 Unique Features of New Design

The Perfostape is a machine tool that can perform the function of both a stapler and a perforator. Shown in Figure 5.3 – 5.6 are some features and componenets of the "Perfostape". The unique features that make it different from the existing perforator and stapler .are herein discussed.

- It is an improved integrated system of a perforator and stapler; constrained yet working independently.
- The perfostape has a universal base onto which both the perforating and stapling units are coupled. Unlike the normal ones that have their individual bases.
- A universal rod connecting the stapler and the perforator end to end acts as a fulcrum were the effect of the turning moment is induced enabling users to perforate and staple with ease
- The stapling pins of the perfostape are to be much larger than that of the ordinary stapler

The diameter of the rods that serve as the paper punch have been increased and also they are made of cast steel which increases the total amount of paper that can be perforated at a time. The presence of the spring forces the pedal to go back to its initial position without any effort from the operator.

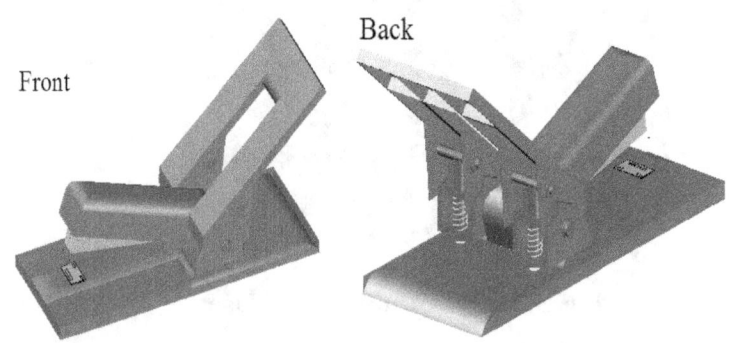

Front   Back

Fig. 5.3: The Perfostape

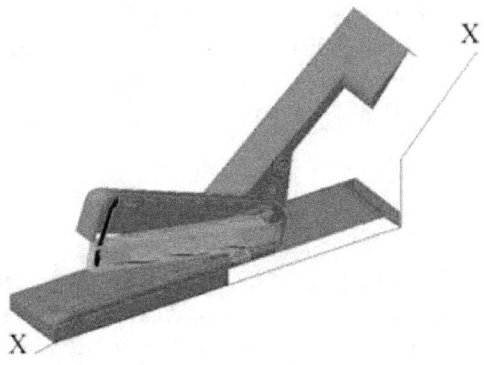

Fig. 5.4: The Sectional View of Perfostapler

5.4 Mechanism of Design

As already mentioned, the perfostape is a simple machine that combines the stapling unit and the perforating unit linked by a fulcrum. The arrangement of the perfostape is in such a way as to allow perforation and stapling simultaneously. There are two major mechanisms involved in its operation, namely: second-order lever mechanism and spring mechanism. A second order lever machine is a simple machine in which the load is between the Effort (i.e. applied force) and Fulcrum. Here the load is the book to be perforated. The arrangement has the stapler unit on the left side and the perforator unit on the right side. This arrangement seems to be the simplest and most feasible.The lever mechanism is discussed below under three sections; namely the stapling, perforating and spring. Stapling: The staple pins are first inserted in the magazine cartridge; the carriage and the tooth are aligned over the cartridge. The Arm is then placed over it. An Effort is applied on the Arm (which is the edge of the perfostape). Perforating: The paper to be perforated is arranged and aligned in the paper slot and an effort is applied.

Spring Mechanism

There are four springs in the perfostape, three are in the stapler unit while one is required for perforation. The springs are called Open-Coiled helical springs.

Spring A: This is found between the cartridge and the base. Its main function is to ensure the return of the magazine cartridge after stapling i.e. it helps the cartridge to be at a particular vertical distance from the base.

Spring B: It is found inside the magazine cartridge and its function is to push against the follow block or pusher. It holds the staple pins firmly and keeps it from shaking or falling, even when it is turned upside down.

Spring C: It is found between the Arm and Tooth. Its major function is to tension the tooth. It pushes the tooth against the pins.

Spring D: This is the only spring used in the perforator unit. It helps to keep the handle and the perforating pins in tension since it is wound around the perforating pins. It also helps to return the perforating pins to its initial position.

5.5 Assembly Guidelines

Based on the exploded view of the perfostape in Figure 5.5 the assembly guidelines are herein presented. The base (B) is attached to the platform (A). The perforating guide (H) is then attached to the base (B). A spring is then attached to the base (B). The magazine cartridge (C) is then placed on the spring. The steel spring (D) is then fixed into the magazine cartridge (C). The carriage (E) is then placed on the steel spring (D). The tooth (F) and arm (G) are then attached together and placed on the carriage (E). The paper punch (K) with the spring (I) attached is then placed in the perforator guide (H). The cylindrical rod (M) is attached to the perforator handle (L) which is then placed in alignment with the perforator guide (H).

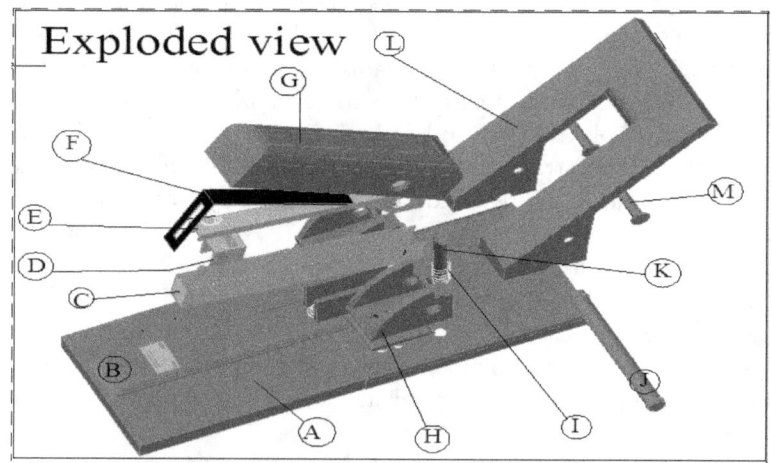

Fig. 5.5: Exploded View of Perfostape

All parts are aligned as shown in Figure 5.10. A fulcrum is slotted in through the perforator guide (H), perforator handle (L), the arm (G) and the magazine cartridge (C).

[6] PORTABLE POSTER REMOVER

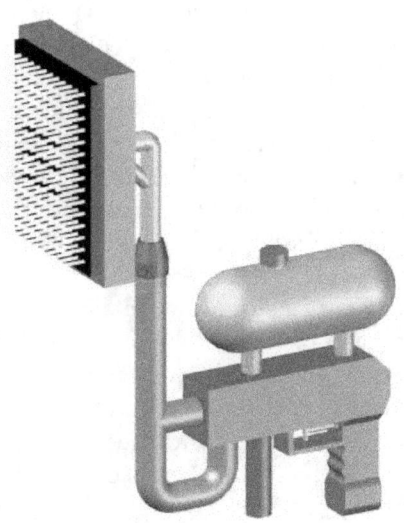

A wall-paper remover with extendable washing arm

"Whatsoever thy hand findeth to do, do it with thy might; for there is no work, nor device, nor knowledge, nor wisdom, in the grave, whither thou goest."—Ecclesiastes 9:10 (KJB).

6.1   Need for Design

The 21st century is one that relies a great deal on the use of posters as a means of information dissemination ranging from burials to crusade to party adverts. These posters are usually attached to notice boards or directly to the wall, but the latter is more common as it is practically impossible to have notice boards all over the place.

The poster has served its purpose, and it's time to remove it, there arises a couple of problem with the removal and restoration of the wall to its original state. The need therefore arises for a device that can remove posters; easily, efficiently and neatly, with minimal stress. This is precisely the purpose of the portable poster remover. It is a simple device consisting of a combination of a solvent for dissolving the paper (and whatever adhesive that has been used) and a soft brush for easy and clean removal of the poster.

6.2   Description of Old Methods

Before now, various methods have been used in removing posters, some just remove the poster with bare hands .This is limited as most time you find bits of the poster still attached to the walls as a result of the adhesive nature, this then gives rise to a second method which is the use of razor blades to scrape off the poster remains. This also has its own limitation as it defaces the wall in the process. A third option has been the use of water and brush. This process is usually tedious and messy. As a result of all these, people have even resorted to contracting the removal of the posters to some laborers. So, these old methods are uneconomical, stressful and messy. Hence the need for a new design that would be portable, economical, neat and also get the work done in the quickest possible time.

6.3   Unique Features of New Design

The new design has a piston which pumps the fluid with less energy. It has a pin which covers the upper outlet and lower inlet of water. It has an adjustable rod which enables it to scrub posters at varying height. The receptacle serves as a temporary storage for the fluid. It has a compact body design that holds the body together as one unit. It has a suitable brush for wall scrubbing has been selected for effective removal of posters without damaging the wall.

6.4   Working Principle of New Design

The description of the new design is done with reference to Figures 6.1- 6.4 The portable poster remover is a device that would easily be used by anybody to get rid of unwanted posters; easily, efficiently and in the least possible time. Fluid is sucked from the tank into the receptacle by pressure applied by pressing the piston. The fluid displaces the pin covering the outlet inlet of the cylinder and then it flows into the cylinder. Force is applied to the piston which drives the fluid in the cylinder to the narrow tube. The force of the fluid displaces the pin from the outlet of the cylinder. The fluid flows into the narrow tube and to the J-shape pipe and to the adjustable pipe. The flow continues until it gets to the brush case and then to the surface of the brush. When the piston is gradually released, the fluid is sucked into the cylinder again while the pin at the outlet closes up. This cycle goes on and on as the piston is pushed in and released.

6.5   Mechanism of New Design

Piston: The piston force pushes the chemical solution to the nozzle when a force is applied to it.

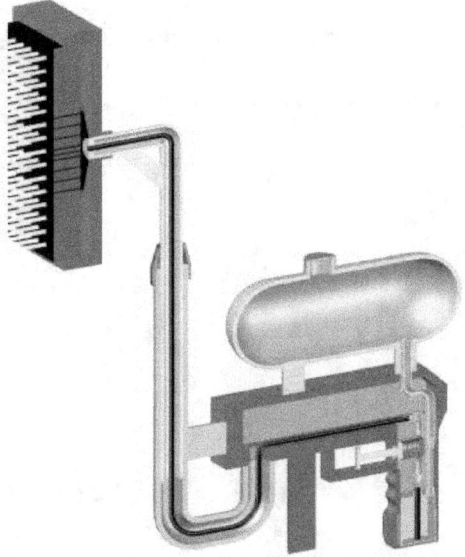

Fig. 6.1: Sectional View of Poster Remover

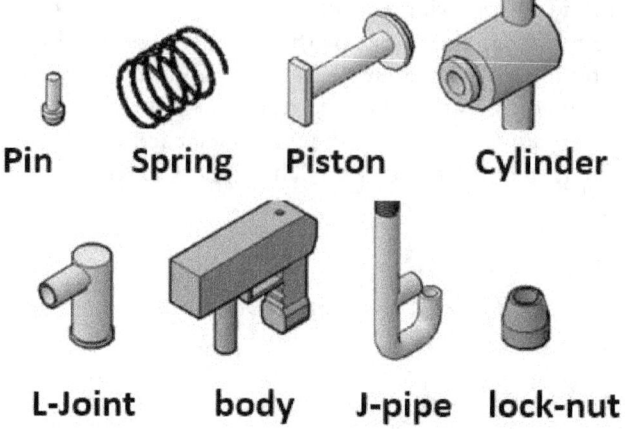

Pin   Spring   Piston   Cylinder

L-Joint   body   J-pipe   lock-nut

Fig. 6.2: Components of Poster Remover

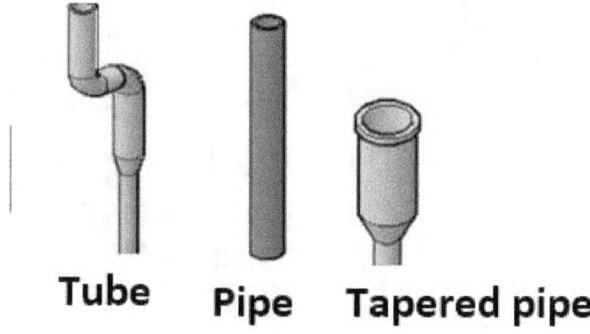

Tube   Pipe   Tapered pipe

Fig. 6.3: **Other Components of Poster** Remover

Pins: There are pins which cover the upper outlet and lower inlet of water. When force is applied to the piston, there is increase in pressure which allows the pin which covers the upper outlet to displace itself and allows the flow of the chemical solution to the nozzle. During this process the swept volume is all covered by the piston. As soon as the piston is released gradually, there is a suction of the solution from the bottom hose which displaces the pin located at the lower inlet to allow the flow of the solution into the vacuum (volume of the cylinder).

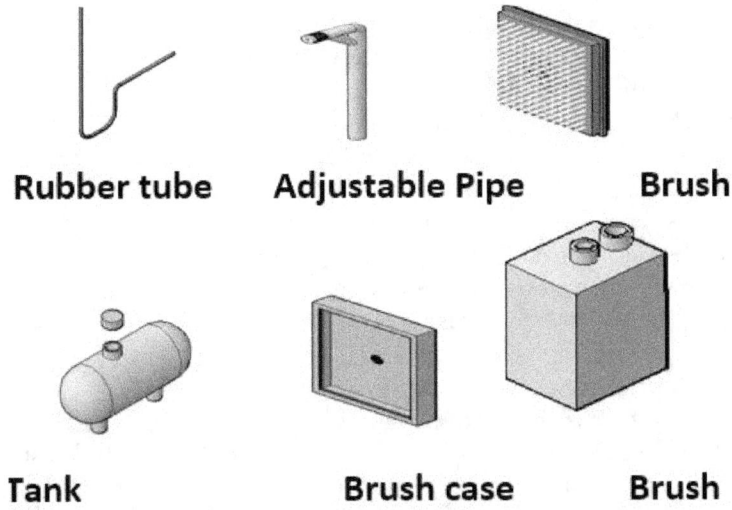

Rubber tube	Adjustable Pipe	Brush

Tank	Brush case	Brush

Fig. 6.4: Other Components of Poster Remover

Adjustable Rod: The rod has been specially designed for comfort-ability while scrubbing. It is an adjustable rod which can be locked by a locknut when the desired scrubbing length is reached.

Receptacle: The receptacle serves as a temporary reservoir for the solution before it is disseminated to the surface of the brush.

Compact Body: The body of the new design (portable poster remover) has been designed, considering an important criterion, compatibility, for easy application/usage of this new machine.

Brush: It has a suitable brush for wall scrubbing which has been selected for effective removal of posters without damaging the wall.

6.6   Assembly Guideline

Based on Figure 6.1 – 6.4, the assembly guidelines are herein presented.

1. The spring sits on the piston and fits on the inner extruded cylinder of the piston.
2. The piston and the spring go into the cylinder.
3. The pins fit into the extruded pipe on both sides (inlet and outlet regions).
4. The L-Joint pipe fits on the upper outlet pipe of the cylinder.
5. The tapered pipe fits on the lower inlet of the cylinder while the pin sits on the lower diameter of the tapered pipe.
6. Straight pipe fits on the lower end of the tapered pipe and its lower end goes into the small receptacle tank which sits on the lower end of the body.
7. The tank suction pipe fits into the second inlet hole of the receptacle tank

and fits into the delivery hole of the tank.

8. The narrow rubber tube is channeled to transfer fluid from the delivery end of the L-joint pipe to the brush case.

9. The piston cylinder arrangement is enclosed in the body

10. The tank sits on the top of the body.

11. The J-shaped pipe fits into the delivery hole of the body through which the tube passes through, from the delivery end of the T-joint pipe to the spray holes of the brush.

12. The lock nut is screwed onto the threaded part of the J-shaped pipe

13. The adjustable rod fits into the hollow region of the J-shaped pipe.

14. The brush case is screwed on the upper end of the adjustable rod.

15. The brush is clamped into the brush case.

6.7   Design Summary

The primary purpose of this work is to design a portable hand poster remover using the mechanism of a toy gun with modifications and addition of brush, adjustable handles, tubes and foam. This objective has been achieved thus far. With the actualization of this design, posters can now be easily removed without defacing the wall and with little energy expended.

PART FOUR

FOOD PROCESSORS

[7] COMPOUND OKRA GRATER

Okra Grating Mechanism

Engineering is the professional art of applying science to the optimum conversion of natural resources to the benefit of man.
---------Ralph J. Smith

7.1   Existing Problems and Need for Design

This project focuses on the improvement of a pre-existing okra grater. An okra grater serves basically to reduce size of an okra pod. This is achieved by moving the okra pod against the grater in a to and fro manner. The application of force on the okra against the grating surface cuts the pods into the required sizes. With the conventional okra grater, grating is done a pod at a time and can only be achieved with the application of appreciable human effort. The aim of the new design is to allow as many pods be grated with the application of the little human effort. There are two essential requirements for a successful grating operation.

- Motion(Translational motion)
- The frictional force
- Pressing force

In the modified design, the translational motion has been converted to rotational motion but the same principle of friction remains the same. The various advantages of the **compound okra grater** over the conventional okra grater will be exclusively discussed in subsequent sections. In addition, there is no contact between the user and the okra pods during the operation. Some of the problems of the pre-existing design are outlined below:

1. It is time consuming; only one okra pod can be grated at a time.
2. The user needs to apply a great deal of pressure manually.
3. The grating surface has sharp protrusions which could be hazardous to the user's hand.

7.2   Description of Existing Design

As shown in Figure 7.1 the existing design is a very simple one made up of; a container, grating surface and a handle.

1. **The Handle -** It is used to hold the grater and the container in place, while grating.
2. **The Grater-** its surface is covered with sharp edged holes used for grating the okra. The grater possesses different sizes of serrations to allow for several sizes of the finished product.
3. **The Container-** It is used to collect the grated okra.

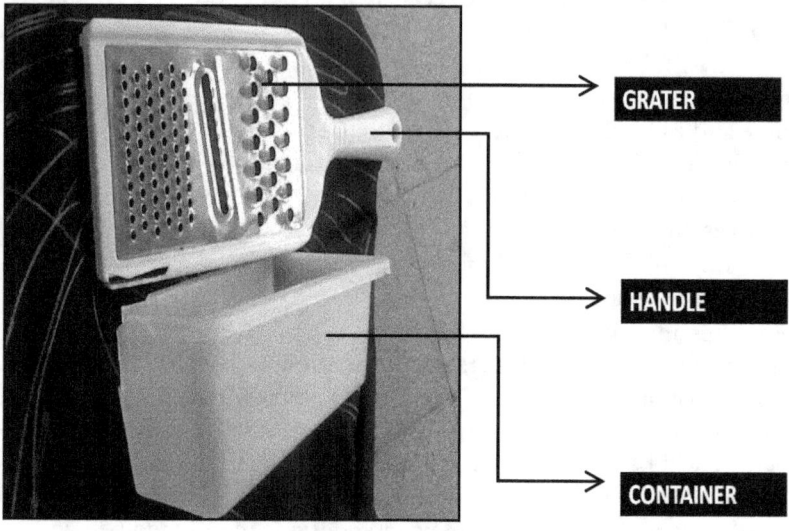

Fig. 7.1: Parts of the Existing Design

7.3   Description of New Design

The new design as shown in Figures 7.2 – 7.4 incorporates the following components which are not part of the pre-existing model.

- The Housing.
- The Cap.
- A steel Weight.
- An Upright handle which is at 90 degrees to the housing.
- Threads which are absent in the old design.

The compound okra grater is basically a grater with sharp-edged slits and perforations on which several Okra pods can be grated simultaneously. It consists of six major parts; Cap; Weight; Handle; Housing; Grater; Container

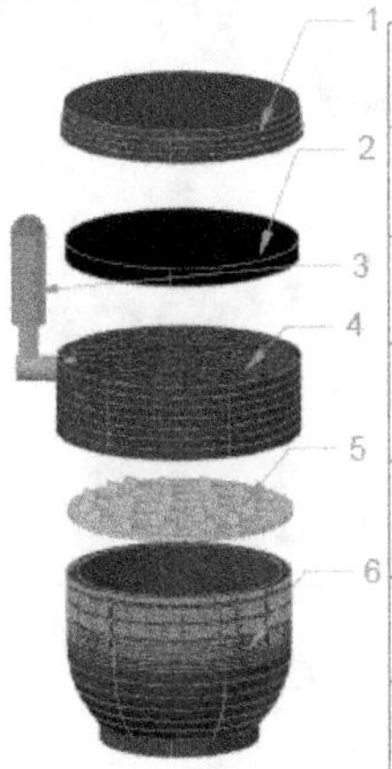

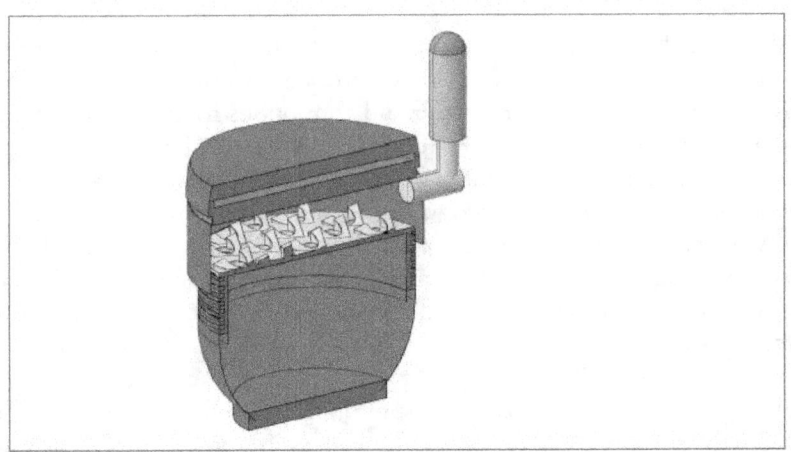

P/NO	PART DESCRIPTION	MATERIAL USED	QTY
1	CAP	PLASTIC	1
2	WEIGHT	STEEL	1
3	HANDLE	WOODEN	1
4	HOUSING	PLASTIC	1
5	GRATER	STAINLESS STEEL	1
6	CONTAINER	PLASTIC	1

Fig. 7.2: Exploded view of Compound Okra Grater

SECTIONED VIEW

Fig. 7.3: Sectioned view of Compound Okra Grater

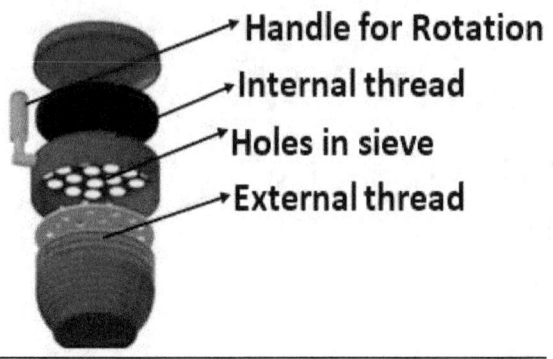

Fig. 7.4: Exploded view of Compound Okra Grater

7.4   Mechanism of the New Design

The grating process is majorly based on a *threading mechanism*. Okra pods are inserted into the holes in the sieve of the *housing* and covered with the *cap* (which houses the steel *weight*). For the okra pods to grate, force as to be applied and this model has been designed to do that with minimal human effort. The force applied to the handle produces a torque that causes the housing to rotate. The inner surface of the housing is lined with *threads* which screw with *external threads* of the *container*. With the aid of the threads, the rotation of the housing causes it to move vertically downwards towards the container.

As shown in Figure 7.4 a *stainless steel grater* is strategically inserted into the top of the container. As the housing is screwed into the container, the okra pods come in contact with the grater and continuous rotation of the housing generates *friction* between the okra pods and the grater. The grater is equipped with sharp protrusions which cut the rotating okra pods as friction is generated. This cutting effect produces the required *size reduction* as is also the case with the conventional okra grater. Continuous rotation of the housing causes the *pressure* exerted on the grater by the pods to increase which in turn causes an upward reaction (Newton 3rd Law of Motion). This reaction is counter-productive as it tends to eject the pods from their slots. To counter this reaction, this design incorporates a mild *steel weight* placed below the cap which is heavy enough to exert pressure on the okra pods and light enough to facilitate smooth rotation of the housing. The presence of this weight ensures that continuous pressure is applied on the grater throughout the operation.

All these in place, the okra pods are grated and pass through *perforations* created by the sharp protrusions on the grater. The grated okra pods fall freely into the container. The operation can be stopped any time in order to view the level or amount of okra grated and at the end of the process, the housing is *unscrewed* from the container to discard the unneeded okra parts and whole process is repeated for new sets of okra pods. The end result of the whole operation is gotten from the container after grating. A stainless steel grater was selected because of its non-rusting property. A wooden handle was chosen over a plastic handle due to its ability to absorb moisture which enhances comfort-ability. Steel was chosen for the weight due to its dense nature and in-activity with moisture. Over all, plastic is preferred over metal due to the fact that after operation, the parts are washed and plastic does not corrode.

7.5 Assembly Guidelines
1. The most important part which is the *grater* should be firmly inserted onto the top of the *container*.
2. The *housing* is screwed on the external threads of the *container*.
3. The *handle* is fit into the body of the *housing*.
4. The *weight* is placed on top of the *housing* to press the okra.
5. The *cap* which covers the *weight* and okra pods fits into the top of the *housing*.

7.6 Design Summary
The Compound Okra Grater is a simple technical model designed to eradicate the problems associated with the conventional okra grater. The introduction of a thread mechanism facilitates an easy and a quick grating operation. With the incorporation of the housing member which can hold as many ten okra pods at once, a much higher productivity rate is guaranteed. The design as a whole facilitates the grating of as many okra pods simultaneously with minimal human effort, at a faster time and less operator risk.

[8] DUAL MODE FOOD MIXER

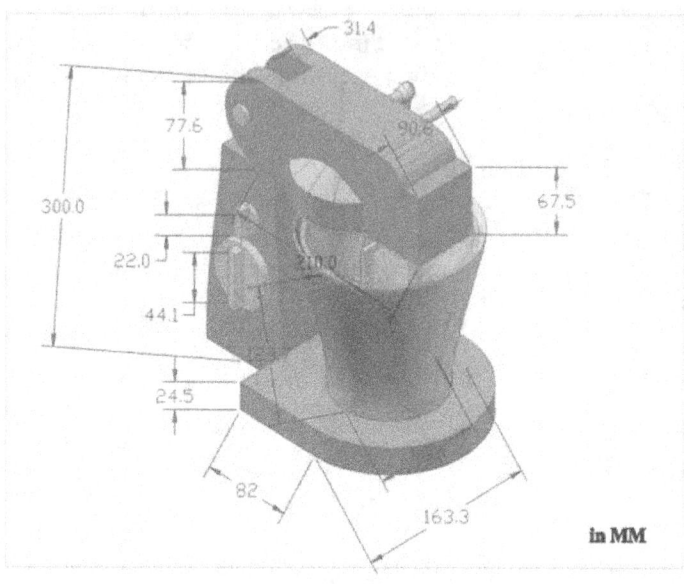

Electrically and hand operated mixer

I wisdom dwell with prudence, and find out knowledge of witty inventions. -------
Proverbs 8:12 (KJB)

8.1 Need for Design

A food mixer is a kitchen appliance used for mixing, beating, and whipping food ingredients. Mixers come in two major variations; hand mixer and stand mixer. A hand mixer is a hand held mixing device. The modern electrically powered type consists of a handle mounted over a large enclosure containing the motor which drives one or two beaters. The beaters are immersed in the food to be mixed. The manually operated type which existed before the invention of the electrically operated type consists of a handle with a hand operated crank on the side. The handle is held with one hand and the crank is turned with the other which drives the beaters. A stand mixer is essentially the same as a hand mixer but is mounted on a stand which bears the weight of the device. Stand mixers are larger and have more powerful motors than their hand held counterparts. They are operated electrically. The Dual Mode food mixer is a stand mixer which can be operated both manually and electrically. It can be used to mix liquid, solid and semi-solid food. It integrates the advantages of the pre-existing designs in one piece and at a reduced cost of production.

The pre-existing food mixer designs were either operated manually or electrically and this posed limitations in the design and usage. The manually operated food mixer in existence today must be used with the two hands at the same time, this makes it difficult to use. The electrically operated one is equally limited because it can only be used when there is power supply and that limits its use to indoors only. The new design became necessary to eliminate the existing limitations since it can be powered both manually and electrically thus making it versatile. It is easy to use and requires little or no skill. It also eliminates the need for each home to purchase both the manual and electrically operated food mixers to satisfy their mixing needs at any time. The dual mode food mixer is designed to make life easy, save money, save time, and give multiple choices to its user. Its main objective is to balance the problem of irregular power supply.

8.2 Description of the Existing Designs

Hand Mixer: As shown in Figure 8.1 and 8.2 the handle (3) is used to hold the hand mixer vertically in place, while the undetectable beater (2) is dipped into the food to be mixed. The handle (1) is then used to rotate the crank (4) which drives the beater (2). The stand mixer above is operated electrically; the plug (9) is inserted into a power supply circuit. The switch button (4) is depressed to on the machine.

Stand Mixer: As shown in Figure 8.2, the speed control knob (7) is used to regulate the voltage supply to the motor assembly (3). The electric motor (3) spindle drives a pair of gears (1), each of which is attached to a beater (8). The turn-table (2) provides a base for the mixing bowl as well as rigidity for the device. The mixer head (10) houses the motor assembly and as well suspends the beater into the mixing bowl. The circuit board (6) is the power unit of the device and supports the speed control switch (5).

8.3 Unique Features of New Design

The features of the dual mixer are discussed in this section with reference to the exploded view shown in Figure 8.3

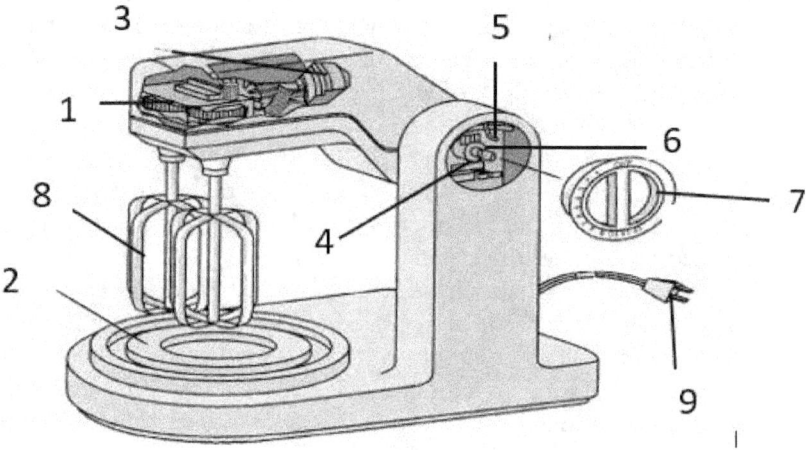

Fig. 8.1: Parts of Stand mixer

1. **The Gear system**: This makes it versatile; it can be used both manually and electrically. This unique addition to the new design is not obtainable in the existing designs.
2. **Detachable beater:** which is spring loaded, hence it can be easily detached or attached without stress. This is not the case in the already existing designs.
3. **The beater:** This is hydro-dynamically designed to reduce fluid friction while in operation, hence increasing efficiency and minimizing power loss. This sophistication makes it better than the existing design.

Fig. 8.2: Modern Stand mixer

4. **Broad base**: This provides rigidity to the device.
5. It has an **Extensible base** which permits different working positions of the mixer bowl. This does not exist in the existing design.
6. It is compact and portable with low noise. This is an improvement on the relatively noisy existing design.
7. The material for the device is specially selected as plastic to ensure light weight.
8. Relatively low cost compared to the existing design.

8.4 Description of the New Design.

The description of this machine is made with reference to Figure 8.3 and 8.4. The dual mode food mixer has a base (4) to sit upon, a bench top and a head (7) hinged to the stand (pillar) (3). The stand (3) is screwed to the mixer bowl base (2) which gives rigidity and support to the mixer. Within the mixer head (7), there is an electric motor (12) which drives the beater (4). The beater (4) is detachable. It can also be fixed to the spur gear (10) hence it can also be driven by the gear system manually. The gear system is made of combination of bevel gear (9) protruded bevel gear (6) and two spur gears (10) i.e. the driver and the driven. The bevel gear (9) is driven by the handle through an effort applied at the handle (13).   The bevel gear (9) in conjunction with the protruded bevel gear (6) helps in transmitting the motion of the handle (13) through a 90 degrees angle. The protruded bevel gear (9) protrudes into the spur gear (10) and drives it along as it rotates. The large spur gear(10) drives the small spur gear and helps to amplify the input effort based on gear ratio, the small spur gear(10) then drives the beater(4).

Also situated in the mixer stand is a switch (16) button which permits the inflow of electric current into the device and supplies the current through the cable (18) to the speed regulator (5). The speed regulator (5) varies the voltage supply of the electric motor, hence the speed of rotation. The bolt (15) and nut (17) are used to join the mixer head (7) to the stand (pillar) (3). The screws (14) are used in holding down the electrical component of the mixer. The vent (11) permits the inflow of air so as to cool the electric components. The mixing head base (8) helps to hold the gear system in place. The mixing bowl l (1) is used to contain the food to be mixed and is placed on the mixing bowl base (2) directly under the beater hence when the mixer is powered either manually or electrically, the food is mixed. Once mixing is completed, the pivot on the head can be raised so as to remove the beater (4) from the bowl. The beater can then be ejected.

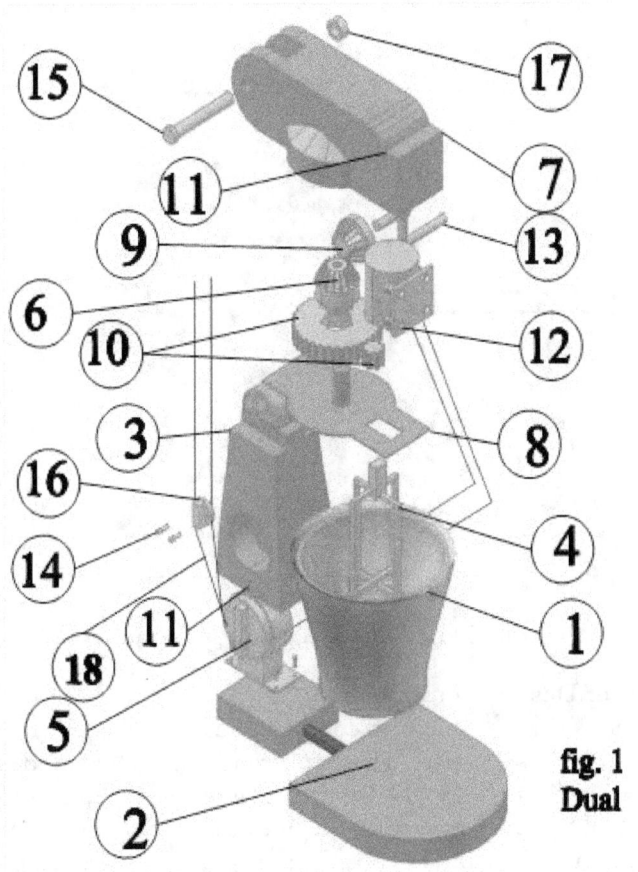

Fig. 8.3: Exploded view of a Dual Mode food mixer.

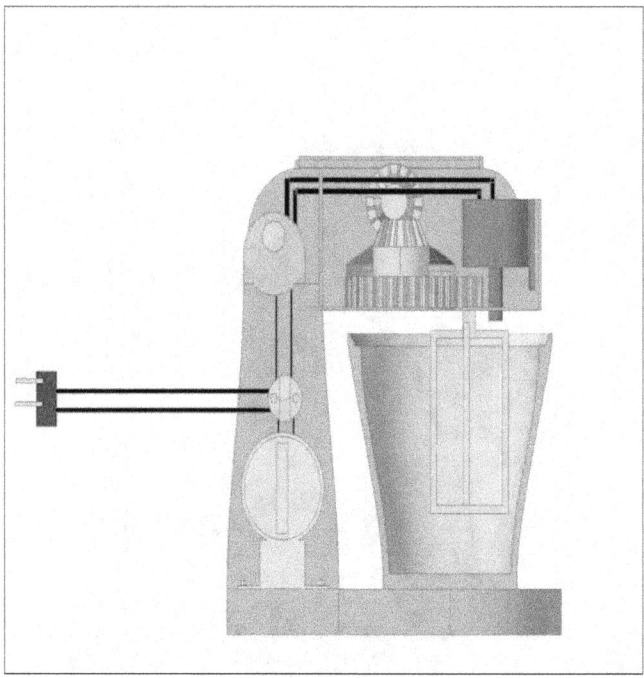

Fig. 8.4: Sectional view of a Dual-Mode food mixer

8.5 Mechanism of Design Components

The Gear System: This is made of a combination of four gears (Fig. 9.5); two bevel gears and two spur gears. They are arranged as shown in the figure below. The bevel gear is driven by the handle and in turn transmits its motion to its pairing mate through an angle of 90 degrees and at the same speed. The protruded bevel gear drives the large spur gear (driver) at the same speed owning to the male and female intersection between them. The large spur gear drives the pinion (driven). The large spur gear has 32 teeth and the small one 8 teeth, hence a gear ratio of 1:4, which means, one revolution of the large spur gear will drive the small gears by 4 rotations and this in turn drives the beater thereby amplifying our input effort.

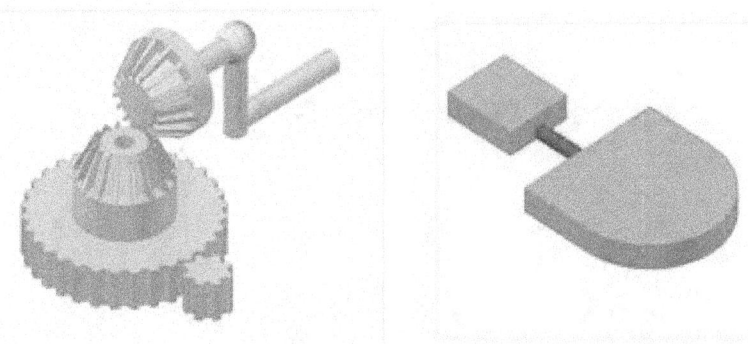

Fig. 8.5: Gear system and extendable bowl base

The Extensible Base: The mixer bowl base is made of two parts connected together by a stud which permits the parts to slide away from each other and brought back to each other whenever necessary. This permits the use of the beater in the manual and electrical positions as the mixer bowl can be kept at different positions.

8.6 Assembly Guideline
1. The cables are connected in this sequence; from the switch button to the regulator and from the regulator to the electric motor.
2. The speed regulator is screwed to the mixer bowl base.
3. The stand (pillar) is screwed firmly to the mixer bowl base to cover the speed regulator.
4. The switch is screwed to button in its position on the stand (pillar).
5. The gears are arranged on the mixer head base as shown in Figure 8.5. The two bevel gears are arranged adjacent to each other, with the protruded bevel gear fixed into the slot on the big spur gear. The big spur gear is then arranged side-by-side with the small spur gear as shown above.
6. The electric motor is fitted to its position on the mixer head.
7. The mixer head is used to cover the mixer head base and screwed in position.
8. The handle is fixed to the bevel gear through a slot on the mixer head.
9. The mixer head is mounted on the stand (pillar) and held in position with the bolt and nut.
10. The beater is fixed either to the small spur gear or the electric motor spindle via a slot on the mixer head base.

[9] TWO-POINT GRINDER

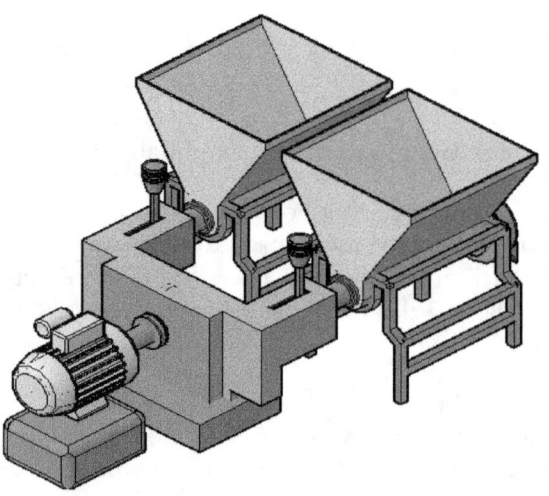

A two-point singular driven grinder

Engineering is the science of economy, of conserving the energy, kinetic and potential provided and stored up by nature for the use of man. It is the business of engineering to utilize this energy to the best advantage, so that there may be the least possible waste. -------William A. Smith

9.1 Need for Design

The need for this novel design stems up from the fact that the operators may sometimes need to grind two different types of food items at the same time. This design is also for large volume of grinding where food is to be prepared for many people within a short time. Moreover, this design will make an addition to the existing food grinding technologies, thus enhancing varieties available for food grinding.

9.2 Description of Existing Grinders

Plate mills as shown in Figure 10.1 are popular in West Africa; they are commonly employed in grinding grains either in wet or dry form. Plate mills operate with a greater component of shear than compression. A plate mill consists of a circular chamber made of cast iron or steel within which two plates with a narrow gap between them are mounted face to face. The plates are grooved in order to provide a shear mechanism. When grains are introduced into the center of the mill, the plates shear the grains between them. One of the plates rotates and the grains revolve, working their way to the outer edge of the plate before dropping by gravity into a holding sack below.

The grains lodge in the rotating plate and are sheared by the grooves in the opposing plate. As the grains move to the edges of the plates, the grooves become shallower and reduce the size of the grains. The design of grooves follows a very old style developed for stone mills several thousand years ago. Plates are usually about 200–300 mm in diameter. Plates are normally aligned in a vertical direction, but horizontal alignment is more convenient when the mill is run by a diesel engine. Plate mills can run as fast as possible but normally at about 2 500–3 500 revolutions/minute, as overheating of the plates limits the speed of the mill.

Frictional heating imposes power limits. For example, a plate mill with 300 mm plates cannot be driven by an engine with more than 12 kW. However, the speed of mill is not a critical factor to the mechanism of grinding. Plate mills operate more effectively with soft and moist grains that shear easily than with hard and brittle grains. It is common in West Africa to add water at the time of grinding. The fineness of the flour ground is adjusted by increasing the pressure on the grain by narrowing the gap between the plates. This is done with a simple hand wheel connected to the outer plate by a shaft. The mill should not be run empty because grains in the mill are needed in order to lubricate the action and, thus, prevent wear. Excessive wear is caused when the plates come into contact with each other. A fine flour or meal from a plate mill is obtained by re-circulating the product in the mill for a second or third grind.

Fig. 9.1: A Fabricated Plate Mill
Source: Ogbe Information.com

9.3 Description of the Two-Point Grinder

Shown in Figures 9.2 – 9.5 are different views of the differential gear-driven food grinding machine. This is a novel design to improve the capacity and universality of the existing food grinders. It enabling the operators to grind large volume of two different food items separately at a time or simultaneously. The machine uses rotational principle of electric motor to supply torque to the differential gear arrangement which uniformly transmit torque through the drive shafts to the mounted grinders (Figures 9.2 and 9.4). The general overview of the machine comprises two major independently integrated systems; the food grinding machine itself as a complete system, and the power transmission (which is the differential gear mechanism) system. However, there are two independently separable food grinders, of equal capacity in the entire system. Each grinder comprises, mainly, the hopper, the grinding chamber, two grinding plates, an auger shaft, frame, support unit, and mechanical devices (e.g. bearings). The power and transmission system comprises the electric motor, the drive/driven shafts, the differential gear arrangement in the gear box, and other mechanical devices. The electric motor converts the energy from the electric power source to the rotational motion on the drive shaft, the drive shaft is the transmission shaft, and as such it transmit power and motion to the gear box housing the differential gear arrangement through the mounted input gear.

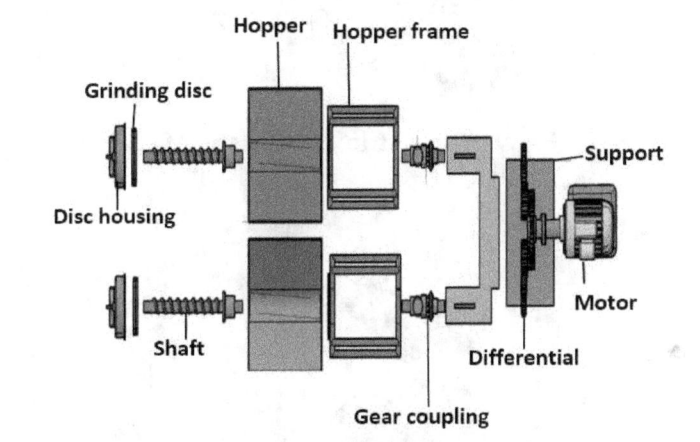

Fig. 9.2 View of the Entire Machine Showing Components

The differential gear arrangement in the gear box simply consists of the spur-gear, which is symmetrical about the central gear that principally and directly receives the drive and torque. It transmits the torque through a couple of meshing spur gears down to the two equal size output spur gears. The output gear meshes during operation with the two output gears. Each of the output gears mounted on the respective feed shaft is connected via coupling system to the auger shaft housed inside the grinding chamber at the lower part of the respective hopper.

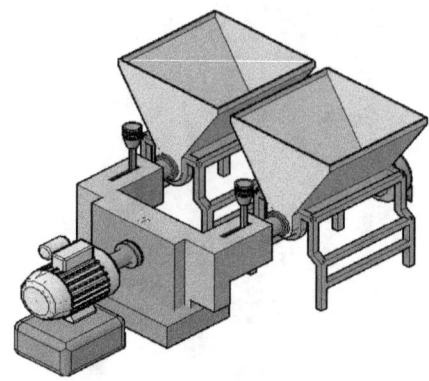

Fig.9.3 Isometric View of the Machine

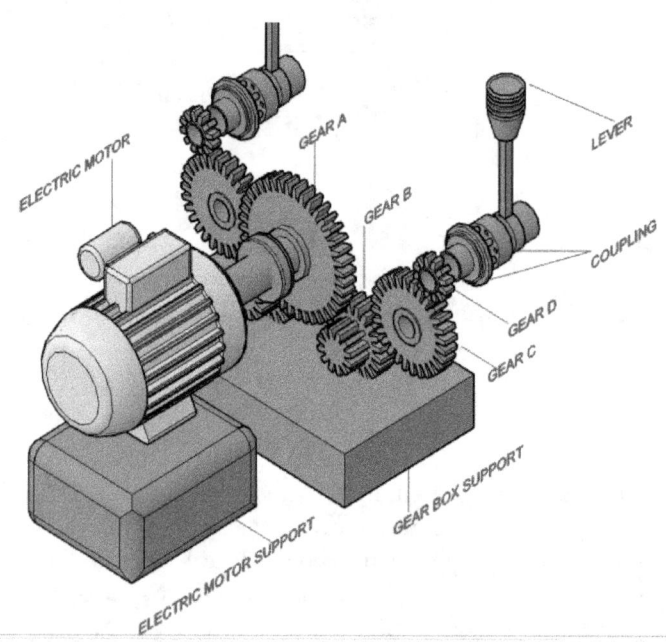

Fig. 9.4: The Power/Drive Transmission Mechanism

The food item to be grinded is fed into the grinder through the hopper; the relative rotational motion of the auger shaft initiates the first grinding impact on the food item. The auger shaft bears a cylindrical spiral profile around a long shaft, this aside from facilitating the crushing of the food item, also paths the crushed item into the secondary grinding chamber. The secondary grinding chamber consists of two circular grinding plates each with a grooved face. The grooved faces of these plates are coupled together inside the grinding chamber. As the food item travels to this chamber, the coupled grooved-face-plates shears the item continuously. A small gate at the lower base of the chamber provides an exit point for the grinded food item which continuously flows into the receiving container

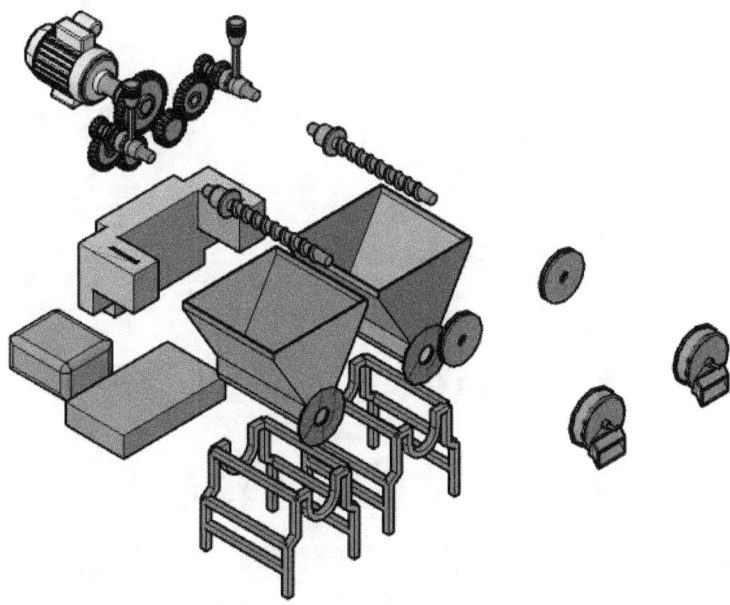

Fig. 9.5 The Exploded View of the Machine

PART FIVE

WORKSHOP
TOOL

[10] MULTI-SPARK PLUG SPANNER

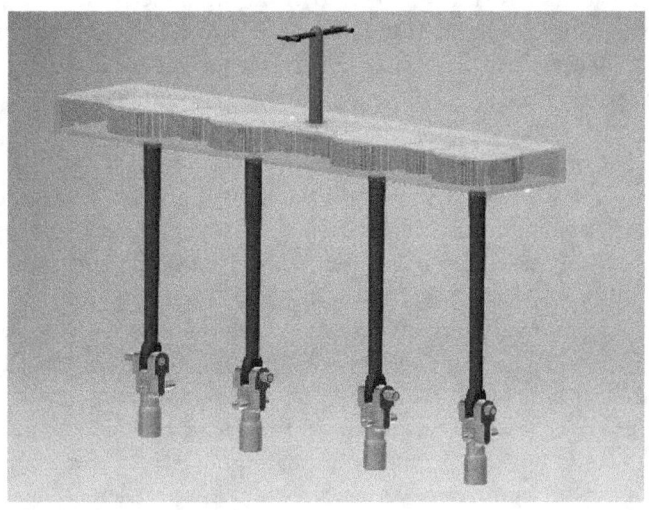

Multi-plug spanner with flexible joints for easy application

Engineering is the professional and systematic application of science to the efficient utilization of natural resources to produce wealth. .....T. J. Hoover and J. C. L. Fish

10.1 Need for Design

The process of removing and tightening spark plugs in vehicle combustion engines is a major activity in any auto mechanic or engine repair workshop. This process is quite tedious and time wasting as the mechanic has no choice than to remove each spark plug one by one. Imagine if all the spark plugs could be removed at once and the time spent on removing all the spark plugs at once is reduced to the time spent to remove just one spark plug. The multi spark plug spanner is designed for an inline 4-spark plug engine.

10.2 Description of Existing Design

The existing spark plug remover (Figure 10.1) consists of a metal handle connected to a socket via the hinges. This socket has the same internal dimensions as the head of a spark plug. The spark plug socket which has a hexagonal mouth is placed over the spark plug and pressure is exerted to be sure that it's all the way down. Torque is applied as shown in Figure 12.1 through the metal ratchet or handle bar to unscrew the spark plug. The socket is rotated in anti-clockwise manner. The problem associated with this mode of removing spark plug is that each of the spark plug has to be removed individually thereby wasting time and creating a stressful repetitive process. Therefore a new design is proposed to solve these problems.

10.3 General and Specific Objectives

The objective is to design a spark plug remover that will be able to remove all four spark plugs of an inline combustion engine all at once. The general objective would be accomplished through the following specific objectives. (i) Design of a rotatable four spark plug remover socket head (ii) Design of gear mechanism to transmit motion (iii) Design of a handle bar for maximum strength.

10.4 Conceptualization of New Design

The new concept of the multi-spark plug remover as shown in Figure 12.2 would have a body made up of four individual spark plug removers. The spark plug removers would be flexible and will also have the ability to rotate by making use of the handle bar which turns the head of each plug. In addition to this, to enable rotation of the spark plug remover mouth, motion is transferred from the head of the socket to the mouth by means of a shaft. Finally to compensate for dimensional differences between engines adjustable hinges will be employed

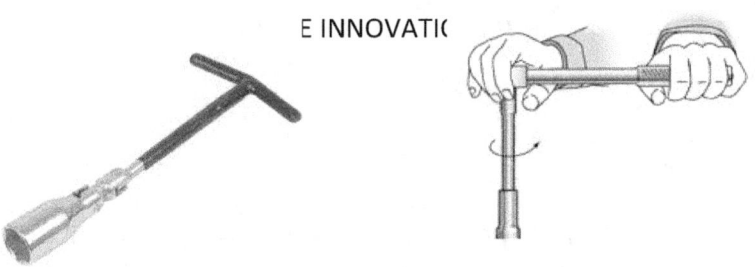

Fig 10.1: Existing Designs of Spark Plug spanners
Source: Toolstation.com

10.5 Description of New Design

The Multi-spark plug remover consists of a combination of nine (9) gears arranged in series. The pinion (shown in Figure 10.2) which transmit the torque, moment and power from the operator, is connected to the other gears so as to deliver this torque to the four sockets attached to the shafts. Spur gears of cast steel material are used in this assembly. The shaft is made with a flat head and connected to a shaft to give a hinge joint. This joint ensures flexibility of device in operation. The shaft connector is held in place with the aid of a pin.

Fig 10.2: A Mesh drawing of gear arrangement

10.6 Description of Parts

Shown in figures 10.3 – 10.11 are the parts of the multi-spark plug spanner.

Pinion: This is that member of the device that is connected to the handle bar. This gear transmits power from the handle bar to the other gears.

Gears: These receive the turning moment from the pinion and transmit the turning moment to the shaft. The gears are also designed with carbon steel material. Each gear is connected to a shaft.

Casing: This houses the gear assembly. It is a hollow box with one hole at the top to connect the pinion to the handle bar and four holes at the bottom where the shafts connect to their respective gears.

Tapered Shaft: This is a solid steel member that connects the gear to the socket. This is adapted with a flat head to connect with a shaft connector, thus forming a hinge joint. This hinge joint is aimed at compensating for differences in inter-sparkplug distances of car engines.

Shaft Connector: This is a high strength, low alloy steel. It connects the tapered shaft to the socket.

Socket: This is that component that fits directly onto the head of the spark plug. It consists of a hexagonal shaped grove which is fixed unto the head of the sparkplug before removal of the plug. It is designed with stainless steel.

Handle Bar: This is a steel member at which the torsional moment is directly applied by the operator.

Handle Bar Shaft: This is a steel member which transmits the torsional moment from the handle bar to the pinion.

Pin: This serves to join the tapered shaft to the shaft connector, and also to join the shaft connector to the socket. It is a spring type straight pins - Coiled, Standard duty ISO 8750 - 8 x 30 pin.

Fig. 10.3: Gear

Fig. 10.4: Pinion

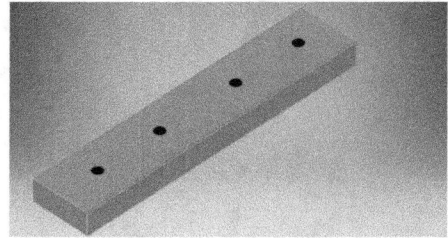

Fig. 10.5: Casing

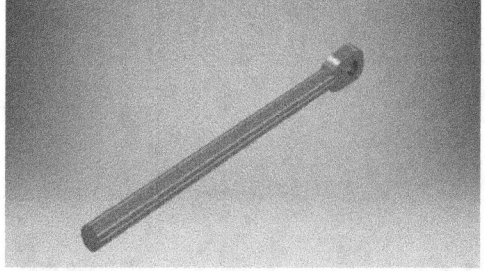

Fig. 10.6: Tapered Shaft

Fig. 10.7: Shaft Connector

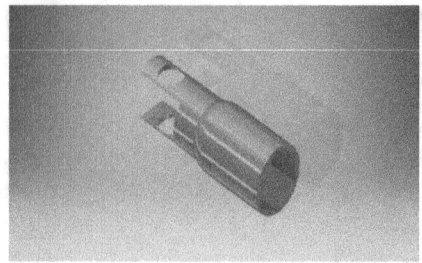

Fig. 10.8: Socket

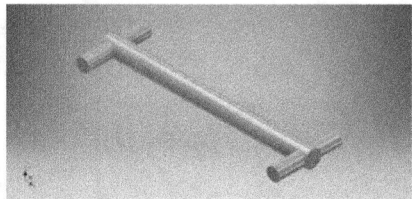

Fig. 10.9: Handle Bar

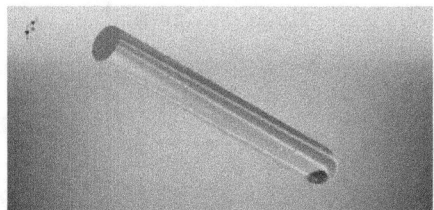

Fig. 10.10: Handle Bar Shaft

Fig. 10.11: Pin

10.7 Mechanism of Operation

The spur gears are arranged side by side as shown in Figure 10.2. This arrangement permits a parallel transmission of power and velocity. The torque supplied by the operator is transmitted to the pinion which is located at the center of the gear arrangement. This pinion serves to drive the remaining gears. Each gear is connected to a tapered shaft, such that the shaft rotates with the gear it is connected to. The gears are arranged such that clockwise rotations of the sockets are achieved when there is a clockwise rotation of the handle bar. Two extra gears are introduced to ensure this.

BIBLIOGRAPHY

A.J. Durelli, E.A. Phillips, C.H. Tsao, Introduction to the Theoretical and Experimental Analysis of Stress and Strain, Mcgraw-Hill Book Company,Inc., 1958.

Allen S. Hall, Alfred R. Holowenko, Herman G. Laughlin, Schaum's Outline Series Theory and Problems of Machine Design

Bill Kerner Ratchets, (assessed 2017), https://www.creativemechanisms.com/ratchets

Bison (Accessed 2017) Machine and bench vises http://www.bison-bial.com/products

Durand, D. 2016 Ratchet Strap Binder and Method of Adjusting a Strap in Length, U.S. Patent No 20160016502 (assessed 2017).

Harman T. L., Dabney J., and Richert N: Advanced Engineering Mathematics, PWS Publishing, Boston, 1997

Higdon A. E., Stiles W. B., and Weese J. A: Mechanics of Materials, 4th Ed., Wiley, New York 1985.

K. Ch. sekhar, V. Jagadeesh, VSSR. Gupta, J. Kalpana, 2017, Design and Fabrication of Manual Peeling Machine. http://www.projecttopics.info/Mechanical/manual-peeling-machine.php

Lee (Accessed 2017) Community Clean http://www.communityclean.co.uk/cleaning/fly-poster-removal.aspx/

Meriam and Kraige, L.G., Engineering Mechanics: Dynamics, 4th ed, New York, 1979

Michael H Bastoni. GEARS Educational systems (Accessed 2017) https://www.manta.com/c/mw2c69g/gears-educational-systems-llc

Peter, W. 2003 Ratchet strap tightener, U.S. Patent No. 6654987B1 [assessed 2017]

PREN 1993-3: 20xx, Eurocode 3: Design of steel structures: Part 1-1: General structural rules, 2001.

R. Gentle, W. Bolton and P. Edwards, 2001 Mechanical Engineering Systems, Butterworth Heinemann, London.

R.S. Khurmi 1995. Theory of Machine. S Chand Publishers New Delhi, India.

Rider Tailgate. Securing a vehicle in a pickup bed or on a trailer using tie-downs assessed 2017), www.ridertailgate.com/securing-vehicle-using-tie-down-straps-s2/

Sarah Shelton Tips on how to use the ratchet [assessed 2017], knowhow.napaonline.com/tips-use-ratchet-straps/

St. Martin's Press 2010, The Bosch book of the Motor Car, Its evolution and engineering development pp. 206-209.

S. P. Ayodeji1, B. O. Akinnuli2, O.M. Olabanji1 2014 Development of Yam Peeling and Slicing Machine for a Yam Processing Plant. Journal of Machinery Manufacturing and Automation . Vol. 3 Iss. 4, PP. 74-83

Willey, Deanna Sclar 2010, Auto Repair for Dummies. Y.C. Fung, Foundations of Solid Mechanics, PRENTICE-HALL, INC.,Englewood Cliffs, New Jersey, 1965.

ABOUT THE AUTHOR

Oluwafunbi Simolowo has been teaching engineering design for over three decades at undergraduate and post graduate levels at the university of Ibadan Nigeria. He has supervised numerous novel graduate and post-graduate projects in the course of his teaching career. He has a number of patents to his credit and was listed as one of his country's (Nigeria) Visionary Innovators on the World Intellectual Property (WIP) day, April 26th 2012. He is at present a Professor at the department of mechanical engineering, faculty of technology
university of Ibadan. Nigeria.

www.ingramcontent.com/pod-product-compliance
Lightning Source LLC
Chambersburg PA
CBHW062247290526
45794CB00006B/2441